Liaison Functions 2022

AA009162

Held at the 2022 AIChE Annual Meeting

Phoenix, Arizona, USA
13-18 November 2022

ISBN: 978-1-7138-7907-7

Printed from e-media with permission by:

Curran Associates, Inc.
57 Morehouse Lane
Red Hook, NY 12571

CURRAN ASSOCIATES INC. proceedings.com

Some format issues inherent in the e-media version may also appear in this print version.

Printed with permission by Curran Associates, Inc. (2023)

For permission requests, please contact AIChE
at the address below.

AIChE
120 Wall Street, FL 23
New York, NY 10005-4020

Phone: (800) 242-4363
Fax: (203) 775-5177

www.aiche.org

Additional copies of this publication are available from:

Curran Associates, Inc.
57 Morehouse Lane
Red Hook, NY 12571 USA
Phone: 845-758-0400
Fax: 845-758-2633
Email: curran@proceedings.com
Web: www.proceedings.com

TABLE OF CONTENTS

AICHE JOURNAL FUTURES: NEW DIRECTIONS IN CHEMICAL ENGINEERING RESEARCH (INVITED TALKS)

ANDREAS ACRIVOS AWARD FOR PROFESSIONAL PROGRESS IN CHEMICAL ENGINEERING LECTURE

EMERGING JUNIOR INVESTIGATOR OPEN INNOVATION FORUM (INVITED TALKS)

JOHN M. PRAUSNITZ AICHE INSTITUTE LECTURE

KICHE-US CHAPTER OPEN FORUM (INVITED TALKS)

SBE'S JAMES E. BAILEY AWARD LECTURE

THE LANGER PRIZE FOR INNOVATION AND ENTREPRENEURIAL EXCELLENCE AWARD PRESENTATION AND LECTURE

WILLIAM R. SCHOWALTER LECTURE

2022 DANCKWERTS LECTURE

IACCHE'S JAMES Y. OLDSHUE LECTURE

FUNDAMENTAL INTERACTIONS OF MICROBES AND MICROBIAL COMMUNITIES WITH MATERIALS

TECHNOLOGIES FOR UNDERSTANDING MICROBIAL INTERACTIONS

PLENARY SESSION: TURBULENCE AND MIXING – IN MEMORY OF PROFESSOR ROBERT BRODKEY I (INVITED TALKS)

69a Superstructure Design of Solvent-Assisted Plastics Recycling Processes

Liela Clarke

Despite the extensive history of plastic manufacturing and widespread use, much of this industry's waste goes unrecycled. The current waste management infrastructure lacks an effective sorting method for plastic waste and as a result a portion of the plastics sent to material recovery facilities are rejected. Conventional mechanical recycling downgrades the recycled plastics properties whereas chemical recycling presents a promising alternative this problem. However, plastic recycling is not economically viable, solvent recovery is a necessary step because of the large volume of solvent required. To that end, difficult problems posed by plastics entering their end-of-life stage need to be addressed through the investigation of feasible chemical recycling methods.
To address this issue, we present a superstructure optimization approach for the recovery and recycling of plastic. With this framework, we categorize and compare different separation technologies based on driving forces, efficiencies, feed properties, and group them into stages. Not only do we consider the technologies used but also the physicochemical properties of the chosen solvent. Within the superstructure, technologies are divided into four stages, namely, impurity removal, recovery, purification, and refinement. Technologies are mathematically modeled in the General Algebraic Modeling Systems (GAMS) software as a mixed-integer nonlinear programming (MINLP) problem. Through the implementation of binary variables, which are used for making a "yes" or "no" decision, we can select an optimal path that simultaneously quantifies the environmental impact, cost, and recovery of the recycling process and meet quality specifications.
To validate the proposed framework, a case study of poly (ethylene terephthalate) (PET) recycling was developed and analyzed. In this case study, ethyl benzoate was used as a solvent for PET dissolution. The proposed process consists of two steps: (1) a dye removal and (2) polymer recovery. PET waste is first subjected to a dye removal step by dissolution through a solvent ethyl benzoate at 120°C. We can simultaneously quantify the potential environmental advantages of the process along with the cost to obtain PET recovery of up to 80% in addition to solvent recyclability benefits.

69b Cellulose nanofibrils-based materials as a substrate for disinfectant wipes

Luke Berger

Disinfectant wipes are a common way to minimize the spread of the virus. They are widely available and their sales have recently surged dramatically. These commercial wipes however have three, often overlooked, issues: 1) they have a relatively limited capacity to absorb alcohol-water mixture upon saturation 2) they dry out very quickly and 3) they end up in landfills. This study investigated a novel biodegradable/disposable substrate for disinfectant wipes composed of cellulose nanofibrils. The novel substrate exhibited high absorption capacity of alcohol-water mixtures. In addition, the substrate significantly reduced the evaporation rate of 70%-isopropyl alcohol compared to Lysol® wipes (60% of alcohol-water mixture evaporated from the Lysol wipes vs only 15% from our material in the same period of time). Furthermore, a disintegration test showed that these substrates are flushable according to the standardized flushability evaluation procedure. The substrates are prepared from renewable and water-soluble materials which reduces the negative environmental impact of the petroleum-based polymer fibers used in commercial products by reducing the waste that ends up in landfills.

69c The Effects of Sulfur Poisoning on CeO2-Based Oxygen Storage Materials

Zexian He

A well-known application of heterogeneous catalysis is the three-way catalyst (TWC), used to eliminate pollutant gases (such as hydrocarbons, carbon monoxide and nitric oxide) from automobile exhaust. However, the performance of TWCs depends on the air-to-fuel ratio that the catalyst "sees". Ceria (CeO2)-based oxygen storage materials (OSMs) are added to TWCs to moderate the effects of changes in the air-to-fuel ratio, so that high conversions can be maintained. However, sulfur, always present in exhaust gas, can deactivate OSMs. The primary goal of our study is to investigate the relationship between physical properties (such as surface area, crystal size) and sulfur deactivation on pure ceria exposed to different temperatures that would change their physical properties, i.e. exposed to air at 500 through 800oC, and ceria-zirconia mixed oxides (CexZr1-x, x from 1 to 0.5, exposed to air at 650oC). X-ray diffraction (XRD) and N2 physisorption were used to characterize the crystal sizes and surface areas of the materials. The oxygen storage capacity (OSC) was measured between 200 and 600oC by introducing a 2-minute CO pulse and measuring the CO2 production, and the degree of deactivation was evaluated by comparing the OSC before and after SO2 adsorption. The degree of deactivation was also evaluated by comparing the total amount of sulfur adsorbed and desorbed, and the decomposition and desorption temperature of sulfur species formed, using temperature programmed desorption (TPD). Understanding the relationship between physical properties and sulfur deactivation can help us design a more effective and durable TWC.

69d Elucidating the factors that lead to plasticization in cellulose derivatives for greener bioplastics

Camille Brule

Bioplastic cellulose acetate (CA) is the most prolific cellulose ester and is utilized for commercial and industrial applications in screens, cosmetics, and textile fibers. Through hydrolysis and other natural environmental factors, CA will convert back to cellulose and acetic acid leaving no harmful impact on its surrounding environment. Depending on the degree of substitution, CA can even be utilized as a naturally biodegradable thermoplastic as a replacement for synthetic plastics. Small molecules called plasticizers are added to CA to modify its thermal, rheological, and mechanical properties, which aids in its processability; however, many common commercial plasticizers are not 'green.' Diethyl phthalate (DEP), for example, comprises 80% of plasticizers used despite its known endocrine toxicity and environmental cost. In order to facilitate the switch to green plasticizer alternatives, the properties of current plasticizers must be determined, in order to ensure that the adoption of green plasticizers does not compromise performance. This research establishes the physical, chemical, and structural characterization of CA films containing known plasticizers, DEP, triacetin, and bisphenol A, as a function of plasticizer concentration and CA acetyl substitution. These characteristics can be used as a benchmark for the design of greener CA plasticizer alternatives.

69e Synergistic Interactions of Nanonet Forming Peptides to Combat Resistance

Samantha Yang

Multidrug resistant Gram-negative pathogens present an increasingly critical healthcare threat. While the intracellular therapeutic targets of multiple antibiotic classes are well conserved between Gram-positive and Gram-negative bacteria, the outer membrane of the latter is highly efficient at denying entry to xenobiotics, thus impeding drug discovery efforts. Our group previously designed a library of synthetic beta-hairpin antimicrobial peptides (AMPs). Some of the potent candidates displayed the unique ability to form on-demand bacteria-trapping nanonets. We hypothesized that mechanical damage to the outer membranes from entrapment by the nanonets would facilitate the entry of putative Gram-positive-selective antibiotics, thus expanding their activity spectrum against Gram-negative pathogens. We successfully demonstrated synergistic interactions between our peptides and a panel of clinically relevant antibiotic classes, with clear distinction in the synergy profile between non-fibrillating and fibrillating peptides. The inclusion of fibrillating peptides at sub-inhibitory concentrations strongly delayed resistance development against the antibiotic, whereas non-fibrillating peptides did not. Findings of strain-specific synergy profiles led us to postulate that intracellular mechanisms were involved. Overall, this work demonstrated the potential of nanonets-forming AMPs as an adjuvant for unlocking antibiotic activity against multidrug resistant Gram-negative bacteria.

69f Plasmonic Copper Nanoparticles for Photocatalytic Dry Methane Reforming: Synthesis, Surface, and Performance

Javiera Cabezas Parra, Emma-Rose Newmeyer, Jamie North, Dayne F. Swearer

Synthesis gas, or Syngas, is in high demand for its wide variety of uses in electricity generation and as a common chemical feedstock. The most environmentally friendly production method is dry methane reforming (DMR) which produces Syngas by reacting methane with carbon dioxide; however, modern DMR requires extremely high temperatures, making it a prohibitively energy intensive procedure. Plasmonic metal nanostructures (PMNs) have garnered considerable research interest due to their potential for sustainable photocatalytic chemistry, with promising evidence supporting copper-based Antenna-Reactor PMNs specifically for photocatalytic DMR (pDMR). It has been widely reported that modifying synthesis methods affects the surface areas, morphologies, and dispersion of supported nanoparticles, and previous research indicates a direct link between such surface structure and the photocatalytic performance of the PMNs. To identify the synthesis-surface relationship, supported copper nanoparticles were synthesized via co-precipitation while systematically varying the Cu-precursor concentration with respect to different metal oxide supports. The samples were screened for Cu^0 loading relative to single and mixed oxide phases through XRD, XPS, and further characterized via temperature-programmed desorption, reduction, and oxidation. The copper-based PMN samples and data were collected into a library complete for eventual comparative photocatalytic performance testing, with the purpose of determining the ideal range of characteristics for an optimal low-temperature, energy-efficient pDMR photocatalyst. The library data will be presented and the ongoing project addressing the research need to comprehensively compare how surface parameters affect pDMR activity will be described, including future work on photoreactor design for catalytic testing.

69g Design and Application of a Small-Scale 3D Printed Crystallization Apparatus

Joseph Kratz

Transition from batch to continuous pharmaceutical manufacturing has received a significant interest in the past decade, primarily due to the advantages that compact continuous systems offer in terms of efficiency, heat and mass transfer control, and in situ control of product attributes via Process Analytical Technologies (PATs). Pharmaceutical batch manufacturing is only feasible on a large scale and is plagued with issues such as high start up costs, poor scale up performance, and being financially prohibitive for manufacturing orphan drugs. This research is focused on developing an inexpensive 3D printed continuous stirred tank crystallizer. The aim is to apply this technology to lab-scale research as well as small scale production of fine chemicals. An MSMPR (Mixed Suspension Mixed Product Removal) type crystallizer has been designed, that has the capability to be a standalone system in a small-scale continuous production process. By reducing the volume to approx. 100 mL, the cost, complexity, and prohibitive nature of early continuous crystallization development can be greatly reduced while still being large enough to accommodate common PAT probes. This can also be an alternative to a larger scale process as a cost-effective solution for small-volume production. With this 3D printed crystallizer, laboratories testing novel compounds have the opportunity to tailor process development to the intended manufacturing scale and the availability of starting material. Instead of focusing on the production of sample material, workflow can be streamlined towards testing, sample analysis and research of other potentially viable compounds.

Machine Learning Models for the Prediction of Henry's Law Constants of Flavor and Aroma Compounds in Bourbon Whiskey

Samuel Haines, Johnathan Graf, Vance Jaeger

Bourbon whiskey is an American distilled spirit derived from corn and grains. Bourbon is aged in new charred oak barrels from which the whiskey gains its distinctive flavor profile. Hundreds of flavor and odor compounds develop from the barrel's wood and char during the aging process. While it would be desirable to experimentally characterize the chemical profile of each batch of whiskey, several practical limitations prevent direct *in situ* measurements on a large scale. Because the liquid is stored in barrels stacked in warehouses over a period of years, data collection is necessarily slow and cumbersome. Opening a barrel offers opportunities for contamination and alters the chemical profile of the barrel's headspace. Small batches of whiskey, that are being used to develop new recipes, are not large enough to consistently sample. For all these reasons, it would be beneficial to develop a minimally invasive technique for probing a barrel's contents and the concentration of flavor and odor compounds over time. We propose that the barrel's contents can be derived by sampling a barrel's gaseous headspace and measuring the samples using gas chromatography. This method minimally perturbs the system while collecting accurate time sensitive data.

By using Henry's law, the barrel's headspace composition can be related to the solvent composition. However, because whiskey contains hundreds of volatile odor and flavor compounds, the experimental measurement of Henry's law constants is difficult and time consuming, especially for compounds that are not commercially available. Moreover, Henry's law constants change with a solvent's temperature and composition (i.e., alcohol proof). To address these difficulties, we have generated a machine learning (ML) model to predict Henry's law constants for organic molecules in water, ethanol, and water-ethanol mixtures based purely on a molecule's chemical structure.

Experimental data was curated from an online database to train and validate several ML models for water-solvated molecules. Results indicate good predictive power of several models with R-squared values greater than 0.8. Because there is no large database of Henry's law constants with ethanol as the solvent, quantum mechanical (QM) calculations were used to predict Henry's law constants and generate a set of theoretical data for water and ethanol as solvents. QM calculations accurately predicted Henry's law constants for water, based upon experimental databases, thus demonstrating the applicability of QM methods in predicting Henry's law constants. Thus, we assume accurate QM-based predictions for ethanol. However, QM calculations take several hours or days to calculate a Henry's law constant for a single compound, and there were several notable outliers in the QM results. Thus it can be beneficial to train ML models that are less prone to the effects of outliers and can predict Henry's law constants in seconds rather than days. ML models for Henry's law constants in ethanol were thus developed using QM data, and R-squared values up to 0.7 were obtained. Water and ethanol models were then be combined to predict Henry's law constants for water-ethanol mixtures such as Bourbon whiskey.

The ML models generated in this project allow for minimally invasive, accurate sampling of whiskey aging. These methods promise to improve how we understand chemical aging processes in the food and beverage industry. We foresee similar methods being applied throughout the food and beverage industry and for the measurement and modeling of aging processes for woods and other plant-based materials.

69h Advancing cell-free glycoconjugate vaccine production

Sarah Sobol

Glycoconjugate vaccines, composed of a pathogen-specific sugar antigen attached to a protein that stimulates the immune system, are an effective method to prevent bacterial infection. Currently, vaccine production uses cells, requiring centralized facilities and distribution under refrigerated conditions. The Jewett Lab has developed a freeze-dried, cell-free system to produce glycoconjugate vaccines at the point-of-care. Within this cell-free glycoprotein synthesis (CFGpS) platform, a specialized protein, known as PglB, transfers sugars to proteins forming the glycoconjugate vaccine. However, three main challenges prevent the adoption of this technology. First, the PglB enzyme is specific to certain sugars and cannot transfer all vaccine-relevant sugar antigens, such as those from diarrheal disease-causing bacteria Shigella flexneri and Shigella dysenteriae. Second, the platform requires an expensive energy source. Third, the platform is not stable at elevated temperatures. To address the first challenge, we hypothesized mutating PglB would expand its transfer capabilities for non-native sugars. Through western blotting, we demonstrated that mutating PglB increased transfer efficiency of sugar antigens from Shigella flexneri 5-fold and Shigella dysenteriae 2-fold compared to wild-type PglB. To address the second and third challenges, we developed a low-cost, thermostable CFGpS reaction formulation by screening low-cost alternative energy sources that could also increase thermostability. Using this formulation, we produced glycoconjugate vaccines against enterotoxigenic E. coli 078 for ~$0.50 per dose after storage for up to four weeks at 37°C. Overall, we advanced cell-free glycoconjugate vaccine production by increasing the number of potential disease targets and developing a low-cost, thermostable formulation for accessible distribution.

126A Data-driven decision making for autonomous materials synthesis

Pragnay Nevatia

Targeted solid-state synthesis plays a major role in driving materials development. However, synthesizing novel compounds with high purity is challenging. Reaction products generally vary with changes in precursors, heating profile, atmosphere, and quality of mixing. Exploring all possible combinations of these variables in a trial-and-error fashion requires many time-consuming and costly experiments, slowing materials discovery. We have therefore developed a more systematic approach to materials synthesis whereby precursors and conditions are suggested to maximize the driving force associated with formation of the target phase. In cases where these experiments fail, our algorithm learns from the synthesis results and updates its ranking of precursors and conditions accordingly using pairwise reaction analysis.

To design and test the decision-making algorithm, we have prepared an unprecedented dataset containing results from 188 solid-state synthesis experiments targeting YBa2Cu3O7, a well-known high-temperature superconductor. With this data, we show that our physics-informed algorithm outperforms state-of-the-art black-box optimization methods. Considering its success, we plan to integrate the decision-making algorithm with an autonomous synthesis platform known as the "A-Lab," where robots will perform experiments in high-throughput to discover new materials at accelerated rates.

126B Nano energetic Materials: Aluminum/Boron Compound Coated with Perfluoro Plasma Nanofilms

Eugene Haley

Aluminum (Al, 31 kJ/g) and Boron (B, 58 kJ/g) are highly energetic metals with numerous applications such as rocket fuels. Nanoparticles are especially beneficial in overcoming kinetic limitations but suffer because a substantial fraction of their mass is already oxidized in the form of a thin native oxide at the particle surface. The oxide layer represents dead weight that decreases the gravimetric energy density of the material and acts as a diffusion barrier that hinders oxidation. One way to minimize these effects is via the synthesis of Al-B compounds. The presence of Al during oxidation leads to an exothermic thermite reaction, which converts boron oxide into highly energetic B by sacrificing less energetic Al. In this study, we develop a facile method to synthesize energetic Al-B composites by annealing a stoichiometric mixture of Aluminum (70 nanometers) and Boron (500 nanometers). We further treat the synthesized composites by plasma deposition of a thin fluorocarbon film in nonthermal plasma. We characterize the oxidation and energy release of the material by Thermogravimetric Analysis (TGA) and Differential Scanning Calorimetry (DSC). X-Ray Diffraction (XRD) and X-Ray Photoelectron Spectroscopy (XPS) are used to study the composition of the material. High-Resolution Transmission Electron Microscopy (HRTEM), Scanning Transmission Electron Microscopy (STEM), and Energy Dispersive Spectroscopy (EDS) are used to analyze the shape, size, and elemental distribution of the synthesized and coated material.

126C Synthesis of Oxygen-Based Ligands for C-H Bond Activation Catalysts

Yasmin Farzan

The cleavage of specific C-H bonds is very useful for many companies including the fuel and pharmaceutical companies. The foremost issue in this research is the strong energy barrier that needs to be overcome to cleave them, meaning that the main ways that have been implemented to overcome this issue involve expensive catalysts such as platinum complexes or energy-intensive processes that require high temperatures. While this process is ultimately effective in its purpose, it does not achieve this transformation in any specific fashion, as it is often uncertain which C-H bond will break first under such extreme conditions, thus leading to many byproducts (which are typically discarded) and waste generation. In this research, we try to find a catalyst that would lower the activation energy of carbon-hydrogen bond cleavage and use earth-abundant, cheap metals instead (e.g., first-row transition metals). The result of this research would be a step towards stopping global warming and increasing the availability of important products such as energy and drugs to many people, especially in developing countries.

126D Oxidation Of Microvascular Carbon/Carbon Composites By Super-Critical Carbon Dioxide Flow

Stanley Williams II

Concentrating solar power (CSP) systems are increasingly becoming a more significant part of the renewable energy grid in the United States. By concentrating sunlight into a single receiver, a working fluid is heated and usually sent to conventional steam turbines to produce electricity. The use of supercritical carbon dioxide (sCO2) Brayton-cycles have been identified as a replacement for conventional steam turbines in CSP systems, providing higher thermal to electricity efficiencies. With this in mind, the direct use of sCO2 as the heat transfer fluid in gas receivers has been proposed. Furthermore, conventional solar receivers are subjected to thermal stresses due to daily startup operations, and thus prone to fatigue failures. Therefore, the use of a novel Carbon/Carbon composite for a modular gas receiver is being investigated due to its lower coefficient of thermal expansion while maintaining high thermal conductivity and mechanical properties. The chemical reaction between sCO2 and the composite, though not significant at low temperatures, is enhanced at the high operating temperatures of the modules. Due to this effect, mass loss of the composite is a concern. In this presentation, a CFD analysis of sCO2 flow in a C/C composite micro-channel is performed with the reaction at the channel's surface to model the surface recession rate of the channel's walls. Using literature kinetic data for C/C composites at varying stages of production together with kinetic data from experiments for composites fabricated in-house, channel recession rates are obtained to propose a predictive model for the composite's mass loss.

126E Guanidinium Polyelectrolytes for the Direct Air Capture of Carbon Dioxide

Brandon Tapia, Xakin Ramirez Isunza, Connor Farrell, Stephen Martin

This research furthers the study of strong selectivity and high sorption capacity amines for the direct air capture (DAC) of CO2 by examining the addition of guanidinium polyelectrolytes to porous activated carbon. Poly(guanidinium styrene sulfonate), made via the reaction of guanidine carbonate with poly(4-styrenesulfonic acid), was impregnated at different loadings into activated carbon. In isothermal sorption testing under 100% CO2 (30 °C, ~40% relative humidity, 1 atm) pure activated charcoal captured between 1.10 and 1.18 mmol/g. The addition of PGSS at 1.0% weight loading increased sorption capacity to 1.25 mmol/g. Past 1.0% weight loading, capacity decreased, hypothesized to be because of sorbent pore clogging due to the polymer addition and supported by SEM images showing a decrease in pore size. Modifications to reduce clogging involved surface treating the support directly with guanidinium and hydroxylating the support via strong acid baths prior to treating with guanidinium. Preliminary results of modified sorbents suggest increased sorption capacities past 1.25 mmol/g. Humidity was not seen to significantly affect sorption as samples in 65% RH had capacities between 90% to 105% that of samples under dry conditions, indicating applicability to DAC. Desorption occurs below 200 °C without requiring vacuum, which suggests reduced energy requirements for sorbent regeneration, increasing net CO2 removal. This work shows that guanidinium polyelectrolytes on activated carbon support represent a potential avenue for DAC due to their increased capacities over pure activated carbon, ability to capture at atmospheric conditions, including humidity, and their lower temperatures for desorption compared to liquid solvent technologies.

126F Quantifying the Rheological Effect on the Capacitance of Carbon Suspensions

Kaua'i Wu

Wind and solar renewable energy sources are becoming competitive options compared to fossil fuels but new innovation of grid-scale energy storage systems are needed to properly implement them. Semi-solid flow batteries (SSFBs), batteries composed of a slurry active material, are a possible solution due to the decoupling of the battery energy capacity and power density. Conductive carbon black has high potential as a slurry material due to its low capital and high surface area for optimizing capacitance. While the rheological and electrical responses of conductive carbon black slurries have been characterized, there has been seldom reports on the two measurements in tandem. To this end, this research focuses on creating a custom-built electrochemical cell capable of performing electrical measurements while a slurry sample is being sheared. The electrochemical cell, built using 3D printing, attaches onto a rheometer, allowing electrical measurements from a VersaSTAT potentiostat to travel through the rheometer. Reproducible rheological data was generated using a solvent trap to minimize evaporation and a pre-experiment procedure used to erase the sample shear history. The capacitance of a carbon black slurry in 35 mg/ml NaCl solution under shear between 0 to 1250 1/s shear rate was calculated utilizing equivalent circuit modeling of Nyquist plots obtained from electrical impedance spectroscopy (EIS). The results showed an increase in capacitance as shear rate increased, which is inline with previous studies. We showcased that high shear rates are necessary for reasonable capacitance to be achieved within SSFB, influencing future design parameters for SSFBs.

126G PEDOT-NHS a Versatile Conjugated Polyelectrolyte for Bioelectronics

Shiv Patel

Organic bioelectronics has recently offered various healthcare technologies such as neural interfacing electrodes, biochemical sensors, and drug delivery devices, with exciting developments on the horizon. Applications in tissue regeneration, for example, allow for the electroactive nature of organic bioelectronic materials to affect cell adhesion, signaling, and fate. Such applications require increased biocompatibility and biofunctionalization, which are limited by the organic semiconductors currently in use. We demonstrate a novel derivative of an organic semiconductor poly(3,4-ethylenedioxythiophene) (PEDOT) that can covalently link to biomaterials and augment otherwise non-conductive materials with enhanced electroactivity. This is accomplished by synthetically modifying PEDOT with N-hydroxysuccinimide (NHS) activated acid groups that target the amine-rich structure of many biomaterials. PEDOT-NHS was characterized demonstrating similar conductivity and electrochemical behavior to the parent polymer without the NHS-incorporated structural units, suggesting that functionalization can be achieved without significantly compromising the electroactivity of the material. PEDOT-NHS and functionalized biomaterials were assessed for biocompatibility, hemocompatibility, and potential for regenerative engineering. The results of this study demonstrate a novel conducting polymer that can augment biomaterials with enhanced electroactivity that may both improve regenerative outcomes and serve as a tool to investigate the mechanistic benefits associated with electrical stimulation of tissue. The outlined approach has potential beyond the functionalization of synthetic biomaterials, to natural decellularized tissue for applications in nerve regeneration and bone growth.

126H Active Fluctuations in a Self-Healing Epithelial Mimetic Build from Biohybrid Components

Andre Gu, Peter Tran, Arthur Prindle, Jan Steinkuehler, Neha Kamat

With the depletion of fossil fuels used for energy production, the search for alternative energy sources is necessary. One promising alternative is biomechanical energy, which is the energy captured from the movement of living organisms. The capture of biomechanical energy from microorganisms is particularly attractive due to their mobility and relative ease of cultivation. One method to capture this microbial energy is to incorporate microorganisms into biohybrid materials, or materials composed of both living and non-living components. Here, we present a biohybrid material comprised of a membrane foam, containing millions of cell-sized compartments, embedded in an aqueous environment. Remarkably, these structures span lateral dimensions of millimeters and exhibit self-reassembling properties, making them reminiscent of an epithelial monolayer. We synthesize these tissue-like structures by producing aqueous microdroplets, promoting surfactant-lipid exchange, and generating droplet networks through a liquid-liquid phase transfer. We characterize these tissue-like structures using a variety of methods and discuss their mechanical properties. Finally, we demonstrate that our tissue-like structures are readily functionalized by cell-free expressed protein or whole cells. We found that the movement of bacteria encapsulated within these structures induces active membrane fluctuations. We further show that these active fluctuations can mobilize microplastic particles deposited on the surface of a biohybrid material. Thus, we introduce a platform for bottom-up assembly of bioinspired self-healing surface coatings that can harness biomechanical energy to impose mechanical forces on its surroundings.

398a Incorporation of Cellulose Nanocrystals and Reactive Surfactants for Improved Pressure Sensitive Adhesive Performance

Caroline Szczepanski, Justin Hamlin, Md Nuruddin, Volodymyr V. Tarabara

Pressure sensitive adhesives (PSAs), which achieve instantaneous adhesion with the application of light pressure, are used in a large range of commodity applications. Aqueous emulsion polymerizations have become a commonly employed synthetic procedure for PSA development, as it is a more sustainable practice that avoids the use of organic solvents and is well-suited to the introduction of additives that can more precisely tune PSA performance. Cellulose nanocrystals (CNCs) have recently become an additive of interest due to their high abundance in nature, biodegradability, and high modulus. However, a persistent challenge with PSAs formed via in-situ emulsion polymerization is surfactant segregation, particularly during adhesive application and drying.

In this work, PSAs enriched with cellulose nanocrystals (CNCs) were synthesized via in-situ emulsion polymerization. In addition to the inclusion of CNCs in the PSAs, the formulations were stabilized with a reactive surfactant (Hitenol AR-1025, AR) to address surfactant segregation and further enhance adhesive performance. The incorporation of both CNCs reactive surfactant lead to a drastic improvement of peel strength, shear strength, and loop tack with peak properties observed at a CNC concentration of 0.75 parts per hundred monomer (phm). A comparative investigation of PSAs utilizing non-reactive (sodium dodecyl sulfate) surfactant revealed that the enhanced performance can be attributed to the synergistic combination of CNCs and reactive surfactant, as only modest improvements can be attributed to surfactant type. In contrast to previous studies that report a trade-off in adhesive properties, this work presents a well-rounded PSA with exceptional peel strength, shear strength, and loop tack.

398b Engineering the Tumor Cell Niche to Study Dormancy in Metastatic Breast Cancer

Shreyas Rao

In recent years, engineered environments have been increasingly employed in cancer research to recapitulate the complex in vivo tumor microenvironment with the ultimate goal of creating a system to better understand disease initiation and progression. In particular, biomaterial scaffolds have emerged as a key tool, enabling design of environments that mimic some features of native tissue and thus provide a controllable but more physiologically relevant environment in vitro. In this talk, I will present our recent efforts in engineering biomimetic environments in vitro using biomaterial scaffolds to study dormancy in brain metastatic breast cancer. Such systems could enable identification of approaches aimed at re-educating the microenvironment with strong potential to halt cancer progression and improve patient survival outcomes.

398c Accelerating the Development of High-Performing Dynamic Electrochemical Processes via Bayesian Optimization

Miguel Modestino, Daniel Frey

The implementation of electrochemical processes to produce fuels, commodity chemicals, and intermediates can allow for the direct integration of renewable energy sources into in major chemical manufacturing processes, and thus help eliminate more than 20% of industrial greenhouse gas emissions. Electrochemical processes provide unique opportunities to increase energy efficiency and enable previously unexplored chemical transformations, but their practical implementation will require the operation of reactors at high selectivity and throughput. Controlling selectivity in electrochemical reactions can be achieved by balancing the mass transport and reaction rates of reactants and intermediates at the electrode/electrolyte interface. This balance can be realized by imposing dynamic potential pulses with tunable magnitudes and timescales to match the requirements of desired reaction pathways. Predicting optimal pulse sequences is challenging because it requires an accurate knowledge of transport properties, reaction mechanisms, and electrochemical rate laws. Alternatively, empirical searches may be used to discover optimal sequences, but the large number of parameters (i.e., multiple pulse times and potential magnitudes) results in an experimentally inaccessible combinatorial design space.

In this presentation, we will discuss how Bayesian Optimization (BO) approaches can be implemented to efficiently explore the design space of pulsed electrochemical processes and lead to enhancement in performance. Electrooxidation of Cerium(III) to Ce(IV) electrolytes will be discussed as a model electrochemical reaction. This reaction is relevant to redox flow battery and redox-mediated water electrolysis, and due to its high oxidation potential, it competes with parasitic oxygen evolution reaction (OER) which lowers its Faradaic Efficiency (FE). We will show how dynamic potential dosing can help balance Ce(III) transport and electrooxidation rates, allowing finer control over its concentration near the electrode and ultimately enhancing FE. Due to the large number of possible pulse sequences, continuum transport and reaction models were built to identify the appropriate window of operating conditions, and then BO was used to rapidly identify the optimal potential pulses. After 25 experiments, a maximum FE = 0.91 was achieved (active pulse time = 5ms, resting pulse time = 136ms, and active pulse voltage of 2.5 V vs. Ag/AgCl) compared to FE = 0.55 at constant potential operation. Furthermore, we will present a multi-objective BO method to identify the pareto conditions that lead to the optimal trade-off between FE and Ce(III) electrooxidation rate – two relevant performance metrics for practical implementation. Similar BO approaches implemented for dynamic CO_2 electroreduction and organic electrosynthesis will also be discussed.

398d Mathematical Modeling of the Effects of Wnt-10b on Bone Metabolism

Ashlee Ford Versypt, Carley V. Cook, Mohammad Aminul Islam, Brenda J. Smith

Bone health is determined by factors including bone metabolism or remodeling. Wnt-10b alters osteoblastogenesis through pre-osteoblast proliferation and differentiation and osteoblast apoptosis rate, which collectively lead to the increase of bone density. To model this, we adapted a previously published model of bone remodeling. The resulting model for the bone compartment includes differential equations for active osteoclasts, pre-osteoblasts, osteoblasts, osteocytes, and the amount of bone present at the remodeling site. Our alterations to the original model consist of extending it past a single remodeling cycle and implementing a direct relationship to Wnt-10b. Four new parameters were estimated and validated using normalized data from mice. The model connects Wnt-10b to bone metabolism and predicts the change in trabecular bone volume caused by a change in Wnt-10b input. We find that this model predicts the expected increase in pre-osteoblasts and osteoblasts while also pointing to a decrease in osteoclasts when Wnt-10b is increased.

398e Grid-Responsive Smart Manufacturing: A Perspective for an Interconnected Energy Future in the Industrial Sector

Blake Billings, Kody Powell

With the growing amount of renewable energy sources, the grid has become responsible for accounting for intermittency and the flexibility needed to utilize dynamic sources. Expensive peaking plants and energy storage systems have been proposed as ways to mitigate those problems. There is a large group of energy consumers that can respond to grid conditions. Historically, these consumers have been residential and commercial users, but with modern innovations and practices, industrial consumers have the potential to become a major player in this space. Grid-responsive smart manufacturing can be used to utilize modern tools in manufacturing innovation as enablers for grid response. These modern tools already exist but are not widely used for industrial grid-side energy management. This paper defines grid-responsive smart manufacturing, identifies five major barriers to its widespread implementation, and portrays the path to getting industrial users to be key players in grid stability and flexibility.

398f Science-Based Design of Experiments and Data Analytics for Molecular-to-Systems Engineering

Alexander Dowling, Jialu Wang, Xinhong Liu, William Phillip, Jonathan Ouimet, Laurianne Lair

Environmental and sustainability global challenges are "wickedly" complex. Identifying impactful solutions for these challenges often requires concerted research efforts that span molecular, material, device, systems, and infrastructure length-scales and transcend disciplines. In this talk, we argue predictive multiscale mathematical models, often grounded in scientific theories, provide principled approaches to realize molecular-to-systems engineering. But the selection, training, and validation of said models in practice are often more of an art than a science, despite the recent advances (and hype?) in data science and machine learning. In this talk, we explore systematic, principled approaches to building science-based mathematical models using several examples from chemical engineering. We focus on opportunities to integrate science-based models, data analytics, and computational optimization (mathematical programming).

Part 1: Motivated by the urgent need for new Li-ion battery recycling technologies, we explore molecular-to-systems engineering of diafiltration membrane separations. We show how superstructure optimization using science-based mathematical models facilitates systematic evaluation of trade-offs in membrane system design and sets quantitative materials property targets. But this analysis is predicated on science-based models for membrane transport. With this motivation, we then share how synergies between data analytics (e.g., dynamic modeling, nonlinear parameter) and transport experiments (e.g., sensor design for time-series measurements) led to the new Diafiltration Apparatus for high-Throughput Analysis (DATA) technique for higher throughput membrane characterization. We show how Fisher Information-based metrics quantify the information gain for additional measurements and facilitate the optimization of experimentation conditions.

Part 2: Science-based (model-based) design of experiments (SBDOE) seeks to maximize the information gained in proposed experiments conditioned on an assumed model by manipulating experimental degrees of freedom. Despite the rich history of SBDOE applications and algorithmic advances for chemical engineering, these methods are often underutilized due to the need to integrate expertise in domain modeling, applied statistics, and computational optimization. To help reduce these barriers, we introduce Pyomo.DOE, a new open-source Python package, for SBDOE. We highlight important model formulations to improve the numeric robustness of SBDOE optimization. In a case study, we apply SBDOE (using Pyomo.DOE) to optimize fixed-bed breakthrough experiments to calibrate to a partial differential-algebraic equation

(PDAE) model for CO_2 adsorption on next-generation metal-organic framework (MOF) materials.

References:

Elvis A. Eugene, William A. Phillip, Alexander W. Dowling (2019). "Data Science-Enabled Molecular-to-Systems Engineering for Sustainable Water Treatment", Current Opinion in Chemical Engineering, 26, p. 122-130. doi: 10.1016/j.coche.2019.10.002

Jonathan A. Ouimet, Xinhong Liu, David J. Brown, Elvis A. Eugene, Tylar Popps, Zachary W. Muetzel, Alexander W. Dowling, William A. Phillip (2022). "DATA: Diafiltration Apparatus for high-Throughput Analysis", Journal of Membrane Science, 641(1), p. 119743. doi: 10.1016/j.memsci.2021.119743

Jialu Wang, Alexander Dowling (2022), "Pyomo.DOE: An Open-Source Package for Model-Based Design of Experiments in Python", AIChE Journal (Early View). doi: 10.1002/aic.17813

Building Microbial Chemical Factories: Design, Assembly, and Engineering of Biological Routes to Chemical Compounds

Kristala Prather

Biological systems have the potential to produce a wide array of compounds with uses that include fuels, materials, bulk chemicals, and pharmaceuticals. Our group is focused on applying principles from metabolic engineering and biocatalysis towards the design and construction of novel biosynthetic pathways for specified target compounds. This "retro-biosynthetic design" approach is aided by advancements in the development of new tools under the umbrella of synthetic biology that facilitate re-engineering of biological systems. As new pathways are designed and constructed, typical challenges such as low product yields and titers can hamper development of commercially-relevant processes. The sheer volume of chemicals that ultimately need to be produced also requires the use of a broader range of feedstocks than those traditional employed in bioprocesses. In this talk, I will review our group's sustained efforts to both produce novel compounds through biological synthesis and develop strategies to address the inherent limitations.

504a Programmable Deformation and Instability of Responsive Hydrogels

Ji-Hwan Kang

Soft materials that respond to their surroundings are poised to become an indispensable element of self-powered smart devices. To achieve proper functionality, these systems often require precise chemical-mechanical coordination between various materials. In addition to exploring new coupling mechanisms, thus, it is crucial to develop prototyping methodologies to ease the integration of a variety of materials, especially those exhibiting emergent behavior across different lengths and timescales.

In this talk, I will describe the rational strategies for shape-morphing soft materials based on responsive hydrogels. First, I will discuss deformation-induced self-folding as a powerful approach to creating complex 3D structures using planar patterning. However, due to the bifurcated nature of the configuration space around the flat state, there is an inherent limitation in terms of robustness. To address misfolding, the pre-biasing mechanism of microscale origami is demonstrated by separating the actuation into two stages: vertex buckling and crease folding. Second, I will discuss the dynamic movement and assembly of thermo-responsive hydrogel nanocomposite films containing photochemically patterned plasmonic nanoparticles that produce light-driven wrinkled structures at an air/water interface. Under patterned illumination, distortion of the phase contact line causes capillary attraction and repulsion between multiple disks, resulting in sustained motion under controlled instability. The design principles of manipulating the specific desired properties of soft active matter using our advanced manufacturing platform will open new avenues for the exploitation of such systems while addressing the current challenges of biomimetic soft materials for biomedical, energy, and environmental applications.

504b Innovating Biomaterials Design for Precision Medicine in Biosensing

Jouha Min

Precision medicine is an emerging approach that matches patients with the most appropriate treatment based on the precise molecular features of an individual patient's disease. Despite huge advances in translational research, diagnosis and treatment continue to pose profound challenges in addressing significant health problems, including cancer and infectious diseases. In this talk, I will discuss a project that demonstrate molecular engineering efforts toward precision medicine. The project focuses on new development in point-of-care diagnosis: multiplexed analyses of multiple analytes predictive of impending sepsis based on hydrogel-based arrays combined with computational holographic imaging method. These studies highlight the potential of integrating chemical engineering, electrical engineering and clinical science approaches to develop advanced diagnostics and innovative therapies for precision medicine.

504c Advancing the Fundamental Understanding of Redox-Driven Separations for Sustainable Water Desalination

Taeyoung Kim, Gowri Mohandass, Weikun Chen, Sitaraman Krishnan

Redox-driven water desalination employs electrochemical reactions that take place at the electrode/electrolyte interface as a means to separate salt from water. Using redox-active ions is an emerging trend that integrates the charge/discharge process of redox flow batteries with salt separation, called redox flow desalination, which takes advantage of energy storage and conversion, opening up greater opportunities to effectively manage the water-energy nexus. In addition, the cell potential can be minimized using a single redox couple in a symmetric system, in contrast with conventional electrodialysis-based approaches, wherein the electrolysis of water utilized to drive salt separation requires relatively high equilibrium and overpotential. Thus, small, modular water desalination systems are possible using redox flow desalination. However, there are technical barriers hindering practical applications, including the electrochemical behaviors of redox-active ions under circumneutral aqueous conditions and their interactions with other components of real source waters, which require fundamental research for the rational design and operation of redox-driven separation systems. This talk presents our current strategies for utilizing redox-active ions to drive salt separation. We have explored several redox-active complexed Fe(II)/Fe(III) species showing the solution composition-dependent electrochemical properties, solubility, and stability, all of which greatly limit the choice of a redox couple to mediate salt separation. One successful ligand used to solubilize Fe(II)/Fe(III) species at neutral pH while maintaining electrochemical activity is citrate, serving as a cost-effective and environmentally-sound alternative to broadly used complexing agents such as cyanide. However, a high redox electrolyte concentration used to overcome the limited electrochemical kinetics of the Fe(II)/Fe(III)-citrate complex adversely affected the selectivity of an ion-exchange membrane separating the redox electrolyte from source waters, thus decreasing the separation efficiency. The same issue was identified in a redox-driven solar desalination process developed to enable energy-neutral production of freshwater. Although the use of I^-/I_3^- simultaneously mediated the sunlight-to-electricity conversion and salt separation in a single device, a small amount of the redox couple was released into product water. Therefore, advancing the fundamental understanding of the electrochemical, physicochemical, and transport properties of redox-driven separation processes is of critical importance to move beyond the early-stage research.

504d Biosponge Polymer Membranes for Capturing Unwanted Toxins in the Body

Hee Jeung Oh

Due to longer life expectancies, the prevalence of age-related diseases is increasing rapidly, and the need for developing biomedical devices that can solve big health problems is similarly greater. Inspired by absorption columns, which are routinely used in industry to remove pollutants from chemical streams, this research describes the design of biomedical devices for capturing unwanted toxins in the body. One significant benefit of using polymer membranes is their tunable binding affinity to target molecules using specific chemical, physical, or biological features. One example is using properly designed polymers to remove cancer chemotherapy drugs that are not taken up by the target tumor during chemotherapy to reduce the drugs' toxic side effects. In the context of reducing the toxicity of chemotherapy, we have designed, built, and deployed porous adsorbers for capturing chemotherapy drugs before they spread through the body.

475a From Silicon to Plastic: It's All About Surfaces, Interfaces and Processing

Elsa Reichmanis

Nanostructured materials play important roles in a wide range of advanced technology applications, stimulating interest in the 'Internet of Things'. In response to that interest, printed, flexible and even stretchable electronics have emerged as low-cost alternatives for devices in applications ranging from energy to health care to security. However, their successful commercialization relies on the design and development of sustainable, robust and reliable materials chemistries and processes. While molecular design plays a significant role in determining materials electronic performance, molecular structure is not the only consideration. Device performance also depends critically on surfaces, interfaces and active material assembly/alignment at many length scales. This presentation will explore how surface chemistry considerations, coupled with interfaces and physical interactions can impact the design and development of advanced new materials technologies for applications ranging from devices to energy storage and conversion.

557a Amyloid Assembly: Unveiling Complexity and Tapping Engineering Opportunities

Jin Ryoun Kim

Aggregation of an amyloidogenic protein, b-amyloid (Ab), into oligomeric and fibrillar aggregates is implicated in Alzheimer's disease (AD). Similarly, formation of amyloid aggregates by a protein, a-synuclein (aS), is a causative factor for pathology of Parkinson's disease (PD). Among different aggregate species, oligomeric forms of Ab and aS are most toxic, causing oxidative stress, inflammation and membrane permeabilization, as well as other detrimental effects. Despite recent progress in AD and PD pathology related to aggregation of Ab and aS, respectively, our understanding of molecular basis of these diseases is still incomplete. One important, yet mostly overlooked, factor is the occurrence of Ab-aS interaction, and consequent effects on aggregation as well as cellular manifestations. In the first half of this talk, I will share our recent findings on complexity associated with Ab-aS interactions and discuss implications of these molecular interactions.

An amyloid protein belongs to a group of intrinsically disordered proteins (IDPs). Due to high structural flexibility, occasionally coupled with functional roles, Nature often uses IDPs as structure-switching biosensors for rapid and specific signaling processes. Thus, IDPs may serve as a new source of proteins that can readily be engineered into molecular sensors with unique sensing behaviors. In the second half of my talk, I will describe our recent engineering effort to create a structurally flexible αS variant, PG65-MIMO (a PG65 variant with Multiple-Inputs and Multiple-Outputs), that rapidly generates distinct fluorescence responses to the three major αS conformers (monomers, oligomers, and fibrils).

Overall, my talk will illustrate important challenges to address to better understand complexity of amyloid diseases and present new engineering opportunities with IDPs.

557b Next-Generation Sulfur Cathode: Electrocatalyst to Accelerate Sulfur Conversion Reaction

Jinwoo Lee

Although lithium-sulfur (Li-S) batteries have been expected as a promising high-energy-density battery system, entangled problems in each cell component hamper their practical use. Especially in cathode, low electronic/ionic conductivity of sulfur based active material, continuous dissolution of polysulfide, and sluggish redox kinetics cause the poor cycle stability and energy density. For the energy density of 500~600 W h kg^{-1}, achievement of high areal sulfur loading and low electrolyte to sulfur (E/S) ratio is needed, but conventional carbon/sulfur composite-type cathodes have limitations to satisfy this requirement. In this seminar, next-generation sulfur cathode with the electrocatalyst accelerating the sulfur conversion reaction is introduced as a powerful platform to simultaneously improve the cycle stability, reversible capacity, and rate capability. This seminar aims to present comprehensive overview of previously conducted studies and future research direction to design the efficient electrocatalyst for sulfur cathode. In particular, strategies to search for new electrocatalyst, to improve the catalytic activity and stability of catalyst, and to fundamentally understand the electrocatalytic trends in sulfur cathode will be covered. We believe development of optimized catalytic materials in cathode with the deep fundamental understanding on electrocatalysis enables realization of high areal sulfur loading and low E/S ratio for high energy density of Li-S batteries.

557c Polyolefins: From Reactors to Catalyst Sites

Kyu Yong Choi

Olefin polymerization is one of the most prominent polymerization processes characterized by its heterogeneous nature of catalytic reactions and various types of continuous polymerization reactors. On a reactor scale, polyolefin reactors exhibit strong nonlinear dynamics due to high exothermicity of the reaction and complex heat removal from both catalyst/polymer particle and reactor scales. The catalytic polymerization, represented by Zieger-Natta and metallocene catalysis, inevitably involve both intraparticle and interfacial mass and heat transfer limitations that affect the polymer yield and resulting polymer molecular weight distribution as well as composition distribution. Optimal grade transition and online model-based polymer properties estimation are also two practically important engineering problems that influence the process economy. In gas phase polymerization of ethylene, which is the dominant industrial process, the formation of fines has been the outstanding problems for years. In this talk, it will be illustrated that the root cause of such fine formation phenomena in gas phase polymerization is primarily due to unfavorable spatial distribution of catalyst components and subsequent premature and uncontrolled particle fragmentation.

557d Synthetic Protein Quality Control to Enhance Full-Length Translation in Bacteria

SangWoo Seo

Coupled transcription and translation processes in bacteria cause indiscriminate translation of intact and truncated messenger RNAs, inevitably generating nonfunctional polypeptides. Here, we devised a synthetic protein quality control (ProQC) system that enables translation only when both ends of mRNAs are present and followed by circularization based on sequence-specific RNA–RNA hybridization. We demonstrate that the ProQC system dramatically improved the fraction of full-length proteins among all synthesized polypeptides by selectively translating intact mRNA and reducing abortive translation. As a result, full-length protein synthesis increased up to 2.5-fold without changing the transcription or translation efficiency. Furthermore, we applied the ProQC system for 3-hydroxypropionic acid, violacein and lycopene production by ensuring full-length expression of enzymes in biosynthetic pathways, resulting in 1.6- to 2.3-fold greater biochemical production. We believe that our ProQC system can be universally applied to improve not only the quality of recombinant protein production but also efficiencies of metabolic pathways.

557e Engineering Polymer Physics and Processing for Advanced Materials

Jay Park

Polymers have played vital roles in nearly every application in the current era; as such, their process-structure-property has been thoroughly investigated for decades. The recent advancement in nano-scale fabrication, additive manufacturing, and wearable technologies have resulted in the discovery and engineering of novel properties that arise from polymer and its composites. The Park research group leads both computational and experimental efforts to harness such properties of polymer/composites to manufacture advanced materials. To this end, this talk will focus on recent efforts by Park research group on three topics: i) smart stimuli-responsive fibers and textiles, and ii) multi-material additive manufacturing. Firstly, self-crimping polyester fibers in a side-by-side configuration based on mismatching modulus and coefficient of thermal expansion (CTE) between the two components are fabricated by bicomponent melt-spinning. Due to the change in crystallinity and its morphology, appreciable change in curvature is observed over the wearable temperature of -20 °C to 20 °C. When non-woven fabrics of the fibers are tested, they show an excellent collective response to temperature. Secondly, a new co-extruded filaments of high/low Tg combinations of PC/ABS for 3d printing is presented. The print parts from these filaments enable thermal annealing above Tg of ABS while maintaining dimensional fidelity of built parts; these yield enhanced interlayer adhesion in print-build orientation to yield close to injection-molded ABS tensile strength.

557f Nanoconfinement-Induced Transitions from Linear to Non-Linear Dynamics of Condensed Matter Ranging from Biological to Geological Ones

Younjin Min

The properties of molecules and macromolecules when confined at nanometer scales differ greatly from their bulk properties owing to the introduction of symmetry breaking, structural frustration and confinement-induced entropy loss in the system. This talk will describe how different degrees of nanoconfinement alter the structural organization and interfacial properties of soft matter ranging from non-biological to biological ones to geological ones. In the first part of this presentation, nanoconfinement mediated intermolecular interactions of two types of ionic liquids and their rheological properties will be discussed in conjunction with the analysis of molecular orientation, organization, and assembly at the interfaces. In the second part of the talk, mechanisms on how biomacromolecules such as silk fibroin proteins self-assemble into hierarchical structures at the multiple-length levels will be explained. Finally, how dynamics and relaxation processes of geocolloids transition to display from linear to nonlinear behaviors as the nanoconfinement proceeds will be discussed with perspectives of colloidal interactions. The presentation will conclude with some outlooks on new fundamental insights for the rational design of novel interfacial materials and also for the assessment of colloidal transport of energy-related contaminants in the subsurface environment.

Exploiting Viruses that Kill and Killing Viruses that Exploit: Some Sweet Science

Jonathan Dordick

Nature has provided us with a blueprint to overcome gaps in our biomedical arsenal, whether for sensing, actuating or providing therapeutic advances. Nowhere is this more evident than in nature's most abundant "life" forms – viruses, including bacteriophages. We have exploited various molecular machines in bacteriophages to sense and/or kill human pathogens, thereby providing new routes to biosensors, therapeutics and disinfectants. The modularity of biological macromolecules drives biological evolution, enables new functions to emerge, and improves existing functions. Unfortunately, viruses often exploit us, as we are now all too aware given SARS-CoV-2 and the tragic loss of life and economic destruction caused by COVID-19. We have once again turned to nature's vast arsenal of biomolecules to target viruses, including coronaviruses, among other human pathogens. This includes the use of the anticoagulant heparin, which possesses extraordinary binding to the SARS-CoV-2 spike protein, and heparin/heparan sulfate analogs and other functional polysaccharides that possess exceptional binding without anticoagulation. In this manner, we have uncovered an Achilles heel of many viral pathogens and identify new routes to bring them under control. We have taken cues from nature to design biological molecules and endow them with unique structural and functional properties. The resulting biomolecular designs enable new uses for functional biomolecules, as well as new bioprocess technologies.

159a 2022 Langer Fellow Presentation: 'Developing novel microbial hosts for waste plastic upcycling'

Tae Seok Moon

Globally, over 300 million tons of plastics are manufactured annually, but only a minor fraction of the waste is recycled. The accumulation of waste plastics in landfills has been considered one of the most challenging environmental issues, emphasizing the pressing need to develop new technologies for recycling or upcycling plastics. Poly(ethylene terephthalate) (PET) upcycling can be achieved by depolymerizing PET into terephthalic acid (TPA) and ethylene glycol (EG) and biologically converting these monomers into value-added products. However, there are only a handful of reports demonstrating microbial strains capable of growing on both TPA and EG generated from PET as sole carbon sources. To overcome this critical limitation, we have performed strain screening to discover a *Rhodococcus* strain (named RPET) that can grow well on the alkaline hydrolysis products of PET as the sole carbon source without any purification step. In this talk, I will discuss the development of a new biotechnological method for converting PET into lycopene. Importantly, the current commercial production of lycopene mainly relies on extraction from tomatoes, deepening the issue of food inequality. This work proposes to solve the issues, including food inequality, sustainable chemical production, and plastic pollution. Notably, this work lays the foundation for a system to upcycle PET to value-added chemicals, which is a critical step toward realizing a fully circular plastic economy.

578a From Particles to People and Complex Organizations

John L. Anderson

In 1989 Russel, Saville and Schowalter published the classic book about the dynamics of Brownian particles suspended in fluids. Bill Schowalter then turned his attention to the dynamics of people by becoming the dean of engineering at the University of Illinois. The jump from modeling inanimate objects to leading human beings is often treacherous, but Bill made it look seamless - which of course it is not. Others who followed him were fortunate to have his advice based on experience, not theory. In this presentation I will discuss leadership of technical organizations and posit that engineering-thinking is an asset. The discussion will also include thoughts on what differentiates engineering from science, and the role of diversity in guiding leaders to "do the right thing".

Working with CO_2

Diane Hildebrandt

CO_2 is produced in processes that convert fossil fuel to energy or chemicals. Reducing CO_2 emissions during the chemical conversions occurring in these processes is at the heart of both chemical engineering and the global challenge of climate change. We can ask why processes emit CO_2 and what the minimum amount of CO_2 a process needs to emit is. This quantification of minimum CO_2 emissions allows actual processes to be systematically compared and improved against that minimum target. It is essential to view the process as a *System*, as the choice of chemistry, catalysts, and operating conditions of the individual unit operations impact the CO_2 emissions of the system. In this talk, I will use XTL as a case study and explain how the gasification - Fischer Tropsch synthesis system can be viewed as a Carnot engine that supplies the work required for the chemical transformations. The reaction temperatures of the gasification-synthesis system determine the efficiency of the Carnot engine and it is these inefficiencies that lead to increased CO_2 emissions from the process.

The concept of the Thermodynamic Attainable Region (AR^T) in Gibbs Free Energy – Enthalpy space is introduced. The insights that can be gained from this approach, both from a broader system's view as well as the more detailed view of unit processes, reactor and catalyst development and operation are explored. The results from the AR^T have guided our development of simple, robust XTL processes which have been demonstrated at various scales.

Looking to the future, CO_2 will be used as a feed for manufacturing commodities, such as detergents, lubricants, and plastics, and chemicals for storage of renewable energy. We can extend the AR^T approach to the chemistry, design, and operation of CO_2 utilization systems.

Nanostructured Polymers for Energy Efficient Devices and Specialty Separation Applications

David Suleiman

Polymers are present in almost every single aspect of our lives. Applications range from common daily uses (*e.g.*, contact lenses, diapers, fabrics, non-stick cookware), to more demanding specialty functions (*e.g.*, artificial hip joints, bulletproof vests, cars, planes etc.). This presentation will describe the synthesis of various nanostructured polymers Specific examples will be presented to describe a variety of hydrophobic and hydrophilic domains (*e.g.*, sulfonated poly(styrene-isobutylene-styrene) grafted with hexyl- and butyl-imidazolium chloride ionic liquids, pentafluoro styrene sulfonated poly(arylene ether sulfone) and others). Some of these polymers will incorporate nanoparticles to create polymer nanocomposite membranes (PNMs), creating a more diverse nanostructure for enhanced separations and more efficient devices. Various materials characterization results will be depicted to describe their nanostructure and their chemical, thermal and morphological behavior. In addition, transport property experiments will be presented to describe the applicability of these materials to energy efficient devices and specialty separation applications. Specific energy efficient applications include: direct methanol fuel cells (DMFC) and organic photovoltaics (OPV). Specific specialty separation efforts include: the inactivation of pathogenic bacteria and viruses, the degradation of emerging contaminants, the separation of urea from water, and water desalination.

67a Design Principles to Establish an Aerobic and Anaerobic Bacterial Community

Jenna Ott, Catherine Day, Anna Hancock, Mohamed S. Abou Donia, Sujit Datta

Bacteria are ubiquitous in both our body and our environment, and most often exist in complex, multi-species communities with varying levels of nutrient availability. One such example is the human gut: both aerobic and anaerobic bacteria live in harmony (or disharmony, in the case of infection) in our gut and experience spatially-varying concentrations of nutrients and oxygen. However, how such spatial variations influence, and are in turn influenced by, the spatial organization of these aerobic and anaerobic communities remains poorly understood. Here, we use a novel experimental system to explore the symbiotic relationship between *Escherichia coli* (a model aerobe) and *Clostridium sporogenes* (a model anaerobe). By linking experimental results to a simple theoretical model, we have developed design principles to create aerobic and anaerobic communities that can stably coexist. Our work thus provides a key step towards engineering 'designer' three-dimensional bacterial communities.

67b Dynamic CRISPR-Based Genetic Programs in Microbe-Laden Hydrogels for Bioproduction

Widianti Sugianto, Benjamin Tickman, Alshakim Nelson, James Carothers

There has been a growing interest in using engineered living materials (ELMs) for on-demand bioproduction as compartmentalized bacteria are reported to sustain long-term metabolic activity[1]. However, the extent to which genetic activity can be sustained in compartmentalized bacteria has not been fully explored. In this work, we applied CRISPR-based expression tools on *E. coli* MG1655 compartmentalized in Pluronic F127-bisurethane methacrylate (F127-BUM) hydrogels to investigate the ability of hydrogel-immobilized microbes in sustaining expression of genetic programs in continuous culture conditions. By directly measuring gene expression dynamics in hydrogels, we found that inducible gene expression is sustained for multiple weeks. This study also reveals that even after 19 days of continuous culture, compartmentalized bacteria retain >50% of their original inducible gene expression capacity. The hydrogel-compartmentalized bacteria also maintain responsiveness to multiple induction cycles, retaining >46% of the original gene expression capacity following a second induction cycle. Using the knowledge of dynamic genetic response in microbe-laden hydrogels, we successfully predicted bioproduction phenotypes arising from inducible CRISPR transcriptional activation (CRISPRa) programs controlling a multi-gene biosynthetic pathway of pteridine-derivatives. This work demonstrates that hydrogel-immobilized microbes can retain expression of CRISPR-based genetic programs for sustained bioproduction, foreshadowing the creation of new generation ELMs for controllable, scalable, and versatile on-demand bioproduction.

Reference:

1. Johnston, T. G. *et al.* Compartmentalized microbes and co-cultures in hydrogels for on-demand bioproduction and preservation. *Nature Communications* **11**, 563 (2020).

67c Interaction of Micro- and Nano-Plastics with Marine Bacteria – What's Happening Under the Tip of the 'Plastic-Berg'?

Tania Silva de Oliveira, Arijit Bose

About 150 million tons of plastic are in the world's oceans currently, and 8 million additional tons of plastics are dumped into the ocean each year. By 2050, the weight of plastics in the ocean will exceed the weight of all marine organisms. This is a highly concerning statistic since plastics can have a half-life of several hundred years. Through ocean action, light and wind exposure these plastic pieces eventually break up into millimeter- and lower-sized objects, known as microplastics. The role of this anthropogenic stressor on marine life must be understood for remediation strategies as well as for policy decisions. There has been extensive publicity due to images of plastic material floating in large patches in the ocean and in fresh water. Since the densities of many commercial plastics (including plastic from natural sources, and those considered biodegradable) are larger than that of water, these materials do not stay on the surface, and sediment deeper into the water column - what is visible is only the tip of the 'plastic-berg'. There are 14 million tons of microplastics on the ocean floor. There they not only encounter fish, but also ubiquitous marine cyanobacteria (CB) that are a critical part of the aqueous ecosystem. We use atomic force microscopy, Raman spectroscopy, and scanning electron microscopy to investigate the physical interaction between one of the most predominant seawater cyanobacteria (*Synechococcus elongatus PCC 7002)* and different plastics. Our results indicate a strong sensitivity to the plastic chemistry, size, surface topology and growth stage of bacteria. The physical, biological, and chemical response of the bacteria to these anthropogenic stressors will impact the fate of both the bacteria as well as the plastic contaminants in ocean water.

67e Adhesion Kinetics of *Staphylococcus Aureus* during the First Stages of Biofilm Evolution

Sarees Shaikh, Patrick Ymele-leki

The ability of most bacteria to grow within structured microbial communities called biofilms is one of the major causes of antimicrobial resistance against antibiotics. Biofilms can withstand antibiotic dosages of up to 1000 times higher than they could resist as planktonic cells and are also associated with biofouling, where microorganisms attach to surfaces and reduce the performance of machines. These aggregates of bacteria are mostly found in wet environments and develop under hydrodynamic stress.

Bacterial adhesion on surfaces is the first step in the biofilm formation process. Thus, it is vital to understand the effect of hydrodynamic forces on the kinetics of bacterial adhesion. This study uses the BioFlux 200 system to investigate bacterial adhesion under fluid flow. *Staphylococcus aureus* planktonic cultures are first grown overnight in TSB broth without dextrose and then diluted in PBS to impede further bacterial growth. Cell adhesion assays are then performed under hydrodynamic stress at 37°C using 48-microwell Low Shear plates (maximum shear stress up to 20 dyn/cm^2) with flow channel dimensions of 75μm depth and 350μm width. The system is connected to an inverted microscope, and images are generated at 5-minute intervals.

Data from the study indicates initial rate of cell adhesion decreases with increasing surface shear stress over the range of shear investigated. At shear stress of 1dyn/cm^2 the average initial rate of cell adhesion observed is nearly 42 cells/min whereas at shear stress of 4dyn/cm^2 the initial rate is halved to 21 cell/min. Further analysis also showed that as shear stress increased, the maximum number of cells that could adhere to the surface decreased significantly. For example, a 2.8-fold reduction in maximum number of cells attached per unit area was observed when the surface shear stress was increased from 1dyn/cm^2 to 4dyn/cm^2 within the flow channel. These findings can help us develop a working kinetic model for bacterial adhesion in the early biofilm evolution stages.

The success of this project may lead to improved understanding and control of bacterial adhesion and biofilm evolution. This may ultimately lead to novel diagnoses and treatment methods for global challenges caused by biofilms.

Funder Acknowledgment(s): This study was supported by an NSF CMMI Award # 2000330 to Dr. Patrick Ymele-Leki.

125a Efficient Bacterial mRNA Sequencing in Diverse Species and Co-Cultures

Kellie Heom, Chatarin Wangsanuwat, Lazarina Butkovich, Michelle O'Malley, Siddharth Dey

Microbes in nature remain one of the largest untapped resources for novel bio-based products, including fuels, antibiotics, and waste treatment solutions. Delineating the transcriptomes of these species with RNA-seq would provide powerful datasets that could be used to characterize the cellular and biochemical functions within these species. However, a major challenge in bacterial RNA-seq is that highly abundant ribosomal RNA (rRNA) molecules must be removed prior to sequencing. If left undepleted, rRNA can consume up to 90% of the sequencing data with uninformative reads, resulting in highly inefficient sequencing of messenger RNA (mRNA), that encodes proteins. This work presents a method, EMBR-seq, which depletes 5S, 16S, and 23S rRNA in prokaryotic species through a combination of selective polyadenylation on mRNA, blocking of rRNA-derived cDNA amplification, and targeted RNase H activity on rRNA. Compared to alternate rRNA depletion methods that employ a large array of oligonucleotides, EMBR-seq employs a few short oligonucleotides per rRNA subunit at strategic locations, thereby making this new technology significantly cheaper and simpler to optimize for any specific prokaryotic species. We demonstrate that rRNA content is depleted down to 3% of sequenced reads in *E. coli*, 1% in *G. metallireducens*, and 9% in *Fibrobacter sp. UWB7*. Additionally, we apply EMBR-seq in co-cultures of *Fibrobacter sp. UWB7* with anaerobic fungi, to efficiently obtain RNA-seq data from mixed prokaryotic-eukaryotic samples and quantify the transcriptional changes compared to monoculture growth. This analysis deepens our understanding of the relationship between two species native to the rumen microbiome and how co-culture conditions induce changes in the bacterial stress response and their cellulolytic activity. Thus, EMBR-seq provides an efficient approach to perform RNA-seq quantification of diverse bacterial systems.

125b Development of a Gut-Inducible Expression Toolkit for Engineering in Probiotic Yeast *Saccharomyces Boulardii*

Deniz Durmusoglu, Nathan Crook

Saccharomyces boulardii is a widely used yeast probiotic, demonstrating effectiveness against various gastrointestinal disorders [1]. Thus, *Sb* is a model for commensal gut fungi and a promising chassis for development of personalized engineered probiotic medicines. Previous work demonstrated that constitutive promoters developed for *S. cerevisiae* are functional in *Sb*, allowing them to be used for *in situ* biomolecule production in the mammalian gut. However, promoters that activate in the gastrointestinal environment are necessary for programming *Sb* with stimuli-responsive functions, thereby enhancing the on-demand activity and fitness of the engineered *Sb* and limiting the off-target activity. In this work, we established a toolkit of native gut-inducible promoters in *Sb*. To do so, we measured the transcriptomic responses of *Sb* under various environmental conditions via RNA sequencing. We employed a stepwise complexity introduction model in our investigations, where *in vitro* cultures of single species (*Sb*) were followed by co-culture with commensal strains, and lastly animal models, including animal model of disease. We identified the genes that were differentially expressed under *in vitro* conditions mimicking the gut such as anaerobic, colon-like media. We also identified the genes that are differentially expressed when another commensal was present (*Clostridioides difficile*) with *Sb* in culture. Lastly, we investigated differentially expressed genes in Sb using several mouse models including germ-free, antibiotic-treated, and *Clostridioides difficile* infection. Overall, this work illuminates the physiology of *Sb in vivo* and enables the development of *Sb* with therapeutic functions that execute under defined *in vivo* conditions.

125c Engineered Habitats for Understanding Fungal Growth in Soil

Yi-Syuan Guo, Gregory Bonito, Scott T. Retterer

Soil microbiomes play a crucial role in the development of the rhizosphere. The dynamic chemical exchange, growth, and migration of members of the soil microbiome and developing plants physically and chemically shapes local microenvironments within the rhizosphere, leading to potential improvements in soil hydrology, nutrient cycling, and general promotion of plant health. However, the specific organizational dynamics and physicochemical drivers of these complex interactions between bacteria, fungi, and the plant host, remain largely unidentified. In this study, we utilized the development of engineered habitats to visualize and quantify the impact of spatial confinement and network complexity on fungal growth. Briefly, such microfluidic environments were fabricated using photolithography and conventional soft lithography methods and assembled to glass slides. Microfluidic networks were designed to be a simplified 2D soil-mimicking media. Networks were generated using a Python script that creates a Voronoi-tessellation of grid points that can be randomly offset, "jiggled" to create geometries with systematically varied complexity. The resulting geometries consist of an array of pores (16 and 30 μm, respectively) connected by throats (10-μm wide). Two general designs were applied featuring (1) regular Cartesian porous media and (2) irregular soil-analog media to distinguish the fungal movements through complex networks. The responses of selected fungal strains isolated from soil (*Linnemannia elongata, Podila minutissima, Fusarium falciforme, Laccaria bicolor,* and *Morchella sextalata*) were recorded using time-lapse microscopy. Rates of hyphal penetration and expansion through well-defined cartesian and 'soil' networks were compared and revealed varied impacts on penetration and expansion rates commensurate with the hyphal size and branching frequency. Notably, pore size also had a significant impact on penetration rates for certain species. Our results demonstrate the utility of using microfluidics to systematically vary key spatial parameters to better understand fungal migration, and point towards the future utility in exploring the drivers of multi-kingdom interactions and rhizosphere microbiome assembly.

125d Mini-Bioreactors for the Study of Pneumococcal Cell-Cell Communication

Corine Jackman Burden, Lydia Eutsey, Frederick Lanni, Shelley L. Anna, Luisa Hiller

Microfluidics allows for real-time single cell visualization and high-throughput experimentation. We have leveraged microfluidics to elucidate the relationship between signaling, population density and cellular heterogeneity in *Streptococcus pneumoniae* (Spn). Our long-term goal is to understand whether and how cell-cell communication in Spn orchestrates its lifestyle switch from a commensal to a pathogen. Our microdroplet-based technology platform consists of a highly monodisperse water-in-oil emulsion. We have successfully encapsulated Spn cells, and their signaling molecules, in these miniature bioreactors. Spn undergoes multiple rounds of division within the droplets, providing an optimal opportunity to quantify the association between cell-cell signaling, cell density, and environmental factors such as nutrient availability, pH, oxygen and temperature. We are investigating the kinetics of signaling of the TprA regulator and its cognate peptide PhrA; this signaling system promotes virulence in animal models of Spn carriage and pneumonia. To monitor PhrA signaling in real time, we utilize a reporter where the promoter of PhrA drives the far-red fluorescent protein, *mCardinal*. We have captured cell-density dependent PhrA signaling in a droplet at a single-cell level. We are currently quantifying the association between cell number and signaling, and its dependence on environmental conditions. Further, we are expanding this system to monitor multiple peptide-dependent signaling systems. In the long term, we hope to shed light on the relationship between the kinetics of Spn cell-cell signaling, environmental conditions, and virulence, and to exploit this knowledge for the development of anti-pneumococcal therapies.

125e Heterogeneous Bacterial Energies As a Bet-Hedging Strategy Against Antibiotics

Pushkar Lele

The exometabolome plays a major role in controlling the formation of bacterial communities and their interactions. To understand how dominant components of the exometabolome shape communities, it is necessary to determine their effects on bacterial physiology. I will discuss how we measure bacterial energies with single-cell assays to quantify aspects of their physiological responses to stressors. Our measurements of the energy source, the proton-motive force (PMF), are revealing fundamental insights into the defense mechanisms employed by native bacteria against invaders as well as antibacterial agents. I will present findings that indicate a broad distribution of PMF levels in bacteria when stressed. Presumably, such phenotypic variability promotes fitness by enabling individual cells to employ heterogeneous survival strategies. Finally, I will discuss our quantitative models that reveal the effect of key exo-metabolites on intracellular protein function and colony formation [1].

[1] Gupta et al, PNAS Nexus, 2022

322b Bits and Pieces of the Turbulence Puzzle: A new length scale, deep structural similarity and a mechanism underlying diverse drag reduction strategies.

Ronald Adrian

Being a multi-scale random process, the length and velocity scales of turbulence are essential parts of the conceptualization and analysis of turbulent fluid flow. Quantities such as the integral length scale, the Taylor micro- scale and the Kolmogorov micro-scale are well defined for homogeneous turbulence, but inhomogeneity, finite domain size and finite Reynolds number introduce additional scales and make interpretation of the classical scale tricky. The vortex micro-scale is a new new quantity that clarifies the link between the micro-scales of Taylor and Kolmogorov at finite Reynolds number and provide a physically reasonable measure of viscous vortex diameters. It's derivation and interpretation are presented. The concept of deep similarity between different types of wall-bounded turbulent flows states that the structures associated with different length scales in seemingly disparate flows occur in analogous ways. This analogy makes it possible to predict the existence and general character of unrecognized scales in one flow based on a more complete understanding of another flow. Lastly, while turbulent drag reduction can be accomplished by many different strategies, i.e. addition of dilute polymer solution, acceleration, riblets and patterns on the wall, transverse wall motion, etc., we have found that there is an underlying mechanism that is common to each of them. This mechanism is the suppression of hairpin packet formation by reducing the amplitude of the wall-normal velocity fluctuations in the buffer layer below the threshold needed for auto generation of packets. Suppression leads to many fewer turbulent stress producing eddies in the logarithmic layer.

322c Conditional Moment Methods for Turbulent Reacting Flows

Rodney Fox, Aziz D. Ilgun, Alberto Passalacqua

In turbulent reacting flows, due to presence of turbulence-chemistry interactions, the source terms in the Reynolds-averaged transport equations are unclosed. A common method for closure in computational fluid dynamics (CFD) applications is a moment-based method. Among them, the conditional moment closure (CMC) is a promising candidate, which solves the conditionally averaged transport equations of the joint probability density function of reactive scalars, while the chemical source terms are written in terms of the conditional scalar means [1]. In our recent study, [2] the idea of the application of Gaussian quadrature-based moment methods (QBMM) to the first-order CMC is exploited by first deriving a closed set of differential transport equations of the higher-order moments of joint mixture-fraction and reaction-progress variable, which are then solved via the conditional quadrature method of moments [3]. To show the effect of degree of correlation to the product yield as well as the application of QBMM-CMC, three-dimensional CFD simulations are run inside a multi-inlet vortex reactor with two-step parallel acid-base liquid phase reaction.

References:

1. R. O. Fox, "Computational Models for Turbulent Reacting Flows", Cambridge University Press, Cambridge, UK (2003).

2. A. D. Ilgun, A. Passalacqua, and R. O. Fox, "Solution of the first-order conditional moment closure for multiphase reacting flows using quadrature-based moment methods," Chem. Eng. J. 405, 127020 (2021).

3. C. Yuan and R. O. Fox, "Conditional quadrature method of moments for kinetic equations," J. Comput. Phys. 230(22), 8216–8246 (2011).

322d Turbulent Velocity and Deformation Fields in Practical Dispersion Devices: Building on what Bob has taught us.

Richard Calabrese

Practical dispersion devices such as stirred vessels, packed columns and high shear mixers rely on energy supplied by a stirrer motor or pump to create the flow and deformation rates required for emulsification and solid-liquid dispersion processes. The deformation fields in these mixers exhibit large spatial and temporal variation so the dispersion capacity of the device depends on both its ability to focus energy and direct material through these high shear regions. Even in the absence of drops and particles, it is imperative to have a fundamental understanding of the velocity and deformation rate fields to aid in device selection and process design. These have been well studied both experimentally and numerically in stirred tank and static mixer geometries but less is known about high shear devices such as rotor-stator mixers and valve homogenizers. Rotor-stator devices present larger experimental and numerical challenges due to close tolerances and high rotor speeds. The highest deformation and energy dissipation rates occur in a narrow clearance region between the rotor and stator, exhibiting well-defined flow structures.

After an introduction referring to stirred vessels, the author will present several studies from his lab on rotor-stator geometries using both experimental (LDA, PIV) and simulation techniques (RANS, LES) to gain a better understanding of the velocity and deformation fields for turbulent water flow. When interpreting results, emphasis will be placed on how probing the underlying fluid dynamics informs analysis of dilute dispersion dynamics. Throughout the presentation, I will share memories from my friendship with Bob Brodkey and how he has influenced and supported my career and ability to reason physically as an educator, researcher and mentor. In addition to personal remembrances, Bob's contributions to the North American Mixing Form (NAMF) will also be noted.

322e Bob Brodkey: A Pioneer, A Super-Mentor, and A Passionate Explorer

Suzanne Kresta

Bob Brodkey was an iconic figure in several academic communities, achieving the rank of professor in 1964 (co-incidentally, the same year I was born) and passionately pursuing the identification of coherent structures in turbulent flows throughout his career.

His engagement with the mixing community was a deliberate act of outreach to a chemical engineering practice community from someone whose work was quite fundamental. This commitment was shaped by the 5 years he spent in industry between his PhD and his academic appointment, and by his early and extended collaboration with Herb Toor.

Bob was beloved as a champion to his junior colleagues, and as a strong advocate for the highest standards of scholarship and academic discussion. This continued well past the time he became a professor emeritus, and his attendance at the APS and AIChE meetings each fall anchored his practice and communities.

While he resisted any suggestion that he take on formal leadership roles, his moral influence and academic mentoring had a substantial impact on several generations of scholars.

This talk will provide some perspective on his contributions to mixing research and practice, with a liberal sprinkling of stories and references to enliven the conversation.

322f Mixing and Packing of Binary Particles

Robert Pfeffer

Three different approaches for mixing binary particles are presented. The first requires mixing coarse and fine silica aerogel particles so that the bulk (or packing) density of the mixture is increased, thereby increasing the insulating properties of the aerogels. This was done by combining a negative pressure (vacuum) with an external force field, such as sound waves. The second application requires uniform mixing and a high packing density of 300 and 30 μm aluminum oxide particles acting as surrogates for explosive RDX particles. This was accomplished by adding moisture to uniformly mix the particles and then removing the moisture to increase the packing density. The third application was to mix 2 different nanoparticles on the nanoscale using a micro-jet assisted nano-fluidized bed.

322g Dr. Bob Brodkey - Teacher, Researcher and Visionary

Kris Lakshmanan

The author had Bob Brodkey as his Ph.D. advisor during the early eighties at The Ohio State University. The research area was using 3-dimensional image processing techniques to capture, measure and model fluid mechanics and mixing. It was some of the early pioneering work in Bob's research interests. The author will summarize how his interactions with Bob during the student days that continued on over the next 37 years. The advice Bob provided from the graduate student days helped the author through his career and family life.

The author will also discuss a couple of other Ph.D. students from that time whom Bob advised, and their interactions prepared them for their academic and research careers.

We all are fortunate to have had Bob as a teacher and researcher and will continue to cherish those memories.

399b Cross-Gradient Transport and Mixing in Turbulent Shear Flows

James Hill, Rodney Fox, Michael G. Olsen

Experiments have demonstrated that in some turbulent shear flows turbulent diffusivity is not a scalar, but significant turbulent transport can occur in a direction not aligned with that of the mean concentration gradient. The experiments took place in a rectangular turbulent channel flow of water with a coflowing spanwise ducted rectangular jet in the channel. A dye was added to the jet so that mixing with ambient fluid occurred downstream of the jet exit. Simultaneous measurements of velocity and concentration were made using PIV and PLIF, which allowed determination of velocity and concentration profiles and various other turbulent statistical quantities. In all cases the downstream turbulent transport was greater than the cross-stream transport in the direction of the mean concentration gradient, in agreement with the measurements of Tavoularis and Corrsin. This was true both for the ducted jet experiments and for cases where the velocities in each stream were the same, the transport occurring in the wakes behind the splitter plates. Additional confirmation showed the same results determined by LES. The cross-gradient transport has been used in modeling turbulent transport and mixing in a conditional moment closure. A simple model and argument supporting this phenomenon has been developed by Blaisdell.

A second experiment using measurements made with stereo PIV and PLIF involved the transport and mixing in a multi-inlet vortex reactor (MIVR). This is a device scaled up from microreactors used to determine rates of fast reactions in order to determine the suitability for large-scale nanoprecipitation applications. A rather flat cylindrical reaction chamber has four horizontal tangential inlets and a vertical withdrawal tube on the cylinder's axis, with fluid dye in alternate inlets. The fluid entering the chamber undergoes a helical path on its way from the inlets until it reaches the vertical outlet with a 5-fold increase in velocity. There appears to be no net shear associated with the transport and mixing, but rather considerable stretching of fluid elements occurs with concomitant increase in local scalar gradients along directions of compressive strain, resulting in rapid mixing, as also suggested by previous DNS results in a study of chemical reactions.

399c On Robert Brodkey, suspending solids and transporting slurries

Harry Van den Akker

Without a proper understanding of the fundamentals of turbulence and fluid-particle interactions, designing or debottlenecking processes involving suspensions and slurries is hopelessly risky. This statement may be a good summary of Robert Brodkey's conviction and his sturdy message to all chemical engineers. In his lifelong career, Robert significantly contributed to advancing the state of the art at the interfaces of fluid mechanics, turbulence, multi-phase flows and chemical engineering. While, typically, Robert harboured a sound suspicion against modern computational tools such as computational fluid dynamics techniques and lattice Boltzmann modelling, these tools do increase our insights in e.g., liquid-solid and other multi-phase flows. Two examples will be highlighted by reviewing recent research results with respect to (a) the role of turbulence in suspending solid particles from the base of a stirred vessel and (b) the role of repulsive forces (electric double layers) in making up the effective viscosity of dense caustic slurries.

399d Effects of Shear Flow Structure on Turbulent Transport and Mixing

Dimitrios Papavassiliou, Oanh Pham, Quoc T. Nguyen

One of the main turbulent flow characteristics is its ability to disperse mass particles and heat in a very efficient manner [1]. In general, the multiple length and time scales that characterize the structure of the flow field drive this dispersion. For anisotropic flow fields, such as those in pipes and channels, the effects of viscosity and advection compete with each other as different flow structures contribute to dispersion depending on the distance from the wall and on the properties of the dispersing material. Robert Brodkey and his students designed elegant experiments that allowed the visualization of flow structures as they convected downstream in a turbulent boundary layer [2-4]. They realized that these structures were associated with high Reynolds stresses that were several times larger than the mean around them, revealing that turbulence is produced and sustained through such flow structures. Computations and experiments over several decades cemented the concept of coherent structures and how they produce turbulence [5], including the development of models for the prediction of turbulence based on very large scale motions in the outer flow [6]. While these coherent flow structures produce turbulence, they play a critical role for the transport of matter and energy in shear flows and they enhance mixing, but not all of them have equal importance [7]. Brodkey's passion to design and conduct careful experiments for understanding the fundamentals of turbulence generation and the analogies between momentum, heat and mass transport has been an inspiration. We have used direct numerical simulations of Poiseuille and Couette flows and Lagrangian computations for the transport of passive scalars in turbulent channel flows to explore the effects of coherent flow structures on turbulent transport [8-11]. In this talk, we will visit the flow effects on the dispersion of passive scalars in view of our simulation results, and will discuss the importance of these concepts in mixing applications [12] and in other cases of turbulence, such as blood flow in medical devices [13].

ACKNOWLEDGEMENTS

The support of NSF under grant 1803014 is gratefully acknowledged as are the use of computing facilities at the University of Oklahoma Supercomputing Center for Education and Research (OSCER) and at XSEDE (under allocation CTS-090025).

REFERENCES

1. Tennekes, H., and J.L. Lumley, A First Course In Turbulence, MIT Press, Boston (1972).Tennekes

2. Corino, E.R., and R.S. Brodkey, 1969, "A visual investigation of the wall region in turbulent flow," J. Fluid Mech., 37, pp. 1-30.

3. Nychas, S.G., Hershey, H.C., and R.S. Brodkey, 1973, "A visual study of tubrulent shear flow," J. Fluid Mech., 61(3), pp. 513-540.

4. Praturi, A.K., and R.S., Brodkey, 1978, "A stereoscopic visual study of coherent structures in turbulent shear flow," J. Fluid Mech., 89(2), pp. 251-272.

5. Smits, A.J., McKeon, B.J., and I. Marusic, 2011, "High-Reynolds number wall turbulence," Annu. Rev. Fluid Mech., 43, pp. 353-375.

6. Marusic, I., Mathis, R., and N. Hutchins, 2002, "Predictive model for wall-bounded turbulent

flow," Science, 329, pp. 193-196, 2010.

7. Eckelman, D.L., and T.J. Hanratty, 1972, "Interpretation of measured variations of the eddy conductivity," Int. J. Heat Mass Transfer, 15, pp. 2231-2239.

8. Srinivasan, C., and D.V. Papavassiliou, 2011, "Direction of scalar transport in turbulent channel flow," Physics of Fluids, 23(11), 115105.

9. Karna, A.K., and D.V. Papavassiliou, 2012, "Near-wall velocity structures that drive turbulent transport from a line source at the wall,' Physics of Fluids, 24, 035102.

10. Nguyen, Q., Srinivasan, C., and Papavassiliou, D.V., 2015, "Flow induced separation in wall turbulence", Phys Rev E, 91, 033019.

11. Nguyen, Q., and D.V. Papavassiliou, 2020, "Using helicity to investigate scalar transport in wall turbulence," Physical Review Fluids, 5(6), 062605m.

12. Nguyen, Q., and D.V. Papavassiliou, 2018, "Scalar mixing in anisotropic turbulent flow," AIChE Journal, 64(7), pp. 2803-2815.

13. Pham O.L., Feher, S.E., Nguyen, Q.T., and D.V. Papavassiliou, 2022, "Distribution and history of extensional stresses on vWF surrogate molecules in turbulent flow," Scientific Reports, 12, 171.

399e Turbulence and Mixing - Honoring Professor Robert Stanley Brodkey

Charles Petty

All turbulent flows are three-dimensional and time-dependent (Batchelor, 1960; Monin and Yaglom, 1965; Brodkey, 1967; Tennekes and Lumley, 1977; Pope, 2000; Davidson *et al.*, 2011, Davidson, 2013). Reactive (and non-reactive) multiphase, multicomponent, and immiscible materials are governed by five equations-of-change: (1) mass; (2) linear momentum; (3) angular momentum; (4) energy; and, (5) entropy (Manninen et al., 1996; Churchill, 2014; Marchisio and Fox, 2013). By hypothesis, all material *transport models* for gases, liquids, and solids are frame indifferent (i.e., *objective*). If the deviatoric component of the Cauchy's stress is not symmetric, then the intrinsic and external angular momentum can be exchanged (Dahler, 1959; Dahler and Scriven, 1961 and 1963).

The Coriolis Theorem predicts that the strain rate is *objective*; and, that the Reynolds "stress" is not. However, the Cauchy stress and the Reynolds "stress" are complementary, but not similar. A Universal Realizable Anisotropic PreStress (**URAPS**) closure predicts that the Coriolis acceleration causes an anisotropic re-distribution of the turbulent kinetic energy among the three components of the fluctuating velocity in a rotating homogeneous decay (Koppula *et al.*, 2009; 2011; 2013). Turbulent flows are commonly decomposed into a Reynolds ensemble average and an instantaneous fluctuating field.

The **URAPS** closure simulates turbulent flows of interpenetrating continua. Beyond the Kolmogoroff scale, the mean flow field is spatially homogeneous and the Reynolds "stress" is spatially anisotropic. The Coriolis Theorem predicts that the material strain rate is frame insensitive (i.e., *objective*); and, most significantly, the Reynolds "stress" is frame sensitive (i.e., *non-objective*).

The Cauchy stress and the Reynolds "stress" are not similar for single phase fluids (or for multiphase fluids). However, for single phase fluids, turbulent dispersion is positive; turbulent dissipation is positive; and, turbulent "production" of kinetic energy may be either positive or negative. For a second-order, irreducible, compressible, Newtonian fluid, the Cauchy stress and the strain rate are symmetric and objective. The phenomenological coefficients for this model are objective scalar-valued functions of the local thermodynamic state and the eigenvalues of the local strain rate. For some materials, the Cauchy stress is asymmetric and objective.

The Reynolds average of the equation-of-change for linear momentum is an exact, albeit unclosed, vector-valued equation for the mean velocity field of a single-phase fluid. Unlike the Cauchy stress, the Reynolds "stress" is a dyadic-valued real, symmetric, non-negative, and non-objective operator for all inertial and non-inertial temporal frames-of-reference. The non-objectivity property of the Reynolds "stress" stems directly from the Coriolis Theorem.

Turbulent flows of multiphase fluids are encountered ubiquitously in industry. The mixture model presented herein will include equations-of-change for mass, linear momentum, angular momentum, energy, and entropy. The multiphase mixture theory is developed by using a phase average followed by a Reynolds average beyond the Kolmogoroff scale, even if the mean turbulent flow is spatially homogeneous. The purpose of this presentation is to extend this observation to a multiphase fluid.

Turbulent multiphase flows are encountered ubiquitously in the process industry. Examples include flows within hydrocyclone separators and fluidized bed reactors (see Cocco, et al., 2014). The potential for simulating these flows has primarily occurred because of advancements in computational hardware and software. However, longstanding deficiencies in turbulence closure models limit the practical utility of computational methods as an enabling technology. This presentation will extend a recently developed Reynolds stress closure for single phase fluids to a Reynolds stress closure for interpenetrating multiphase fluids. The normalized Reynolds stress is a non-negative operator; therefore, it is essential that all of the eigenvalues of this operator must be non-negative for all turbulent flows in rotating and non-rotating frames-of-reference. This fundamental mathematical property cannot be compromised if turbulence modeling is to attain its full potential in predicting flows in complex geometries. Although current CFD technology can reproduce benchmark flows, the ability to predict low-order statistical properties beyond a calibrating flow is not possible. This weakness of turbulence modeling can be traced to the closure hypothesis that velocity fluctuations are objective vector fields. This assumption is embedded in the sub-grid closures associated with large-eddy simulations; the pressure/strain rate closures associated with the second-order moment equation for the Reynolds stress; and, the eddy viscosity models associated with the closure of the Reynolds averaged Navier-Stokes equation.

The *ad hoc* assumption that the Reynolds stress is an objective operator similar to the Cauchy stress in continuum mechanics is not supported by direct numerical simulations of the Navier-Stokes equation and fundamental physical principles of thermodynamics and turbulence. Continuum scale hydrodynamic fluctuations, unlike molecular scale fluctuations, are not objective vector fields. Research at Michigan State University has identified a class of algebraic closure models for the normalized Reynolds stress that are realizable for all turbulent flows in rotating and non-rotating frames. This discovery has much potential to transform current CFD technology from an interpolating tool to a predictive tool.

Citations

Batchelor, G. K., 1960, The Theory of Homogeneous Turbulence, Cambridge University Press.

Brodkey, R. S., 1967, The Phenomena of Fluid Motions, Addison-Wesley Pub. Co.

Churchill, S.W., 1996, Critique of Predictive and Correlative Models for Turbulent Flow and Convection, Ind. andamp; Eng. Chem. Res., 35, 3122.

Cocco, R., Kari, S. B. R., Knowlton, T., 2014, "Introduction to Fluidization: Back to Basics", CEP, AIChE.

Dahler, J.S., 1959, Sept. 29, Transport Phenomena in a Fluid Composed of Diatomic Molecules, The Journal of Chemical Physics, Vol. 30, No. 6, pp.1447-1475.

Dahler, J.S. and L.E. Scriven, 1961, Angular Momentum of Continua, October 7, Nature.

Dahler, J.S. and L.E. Scriven, 1963, Theory of Structured Continua. I. General Consideration of Angular Momentum and Polarization, Proceedings of the Royal Society of London, Series A, Mathematical and Physical Sciences, Oct. 29, Vol. 275, No. 1363, pp.504-527.

Davidson, P.A., Y. Kaneda, K. Moffatt, K. R. Screenivasan, 2011, A Voyage Through Turbulence, Cambridge University Press

Davidson, P.A., 2013, Turbulence in Rotating Stratified and Electrically Conducting Fluids, Cambridge University Press.

Koppula, K. S., A. Bénard, and C. A. Petty, 2009, "Realizable Algebraic Reynolds Stress Closure", Chem. Eng. Sci., 64, 4611-4624.

Koppula, K. S., A. Bénard, and C. A. Petty, 2011, "Turbulent Energy Redistribution in Spanwise Rotating Channel Flows", Ind. Eng. Chem. Res., 50 (15), 8905-8916.

Koppula, K.S., S. Muthu, A. Bénard, and C. A. Petty, 2013, "The URAPS closure for the normalized Reynolds stress", Physica Scripta, T155.

Manninen, M., V. Taivassalo, S. Kallio, and, A. Akademi, 1996, On the Mixture Model for Multiphase Flow, Espoo, Technical Research Center of Finland, VTT Publication, 288, 67 pages.

Marchisio, D.L., and R.O. Fox, 27 May, 2013, Computational Models for Polydispered Particulate and Multiphase Systems, Physics, Cambridge University Press.

Monin, P.K., and A. M. Yaglom, 1965, Statistical Fluid Mechanics: Mechanics of Turbulence, Volume 1, The MIT Press.

Pope, S. B., 2000, Turbulent Flows, Cambridge University Press.

Tennekes, H. and J.L. Lumley, 1977, A First Course in Turbulence, The MIT Press.

399f Laminar Chaotic Mixing in 3D Systems - A Thirty-Year Conversation with Robert Brodkey

Fernando Muzzio

When I became friends with Bob Brodkey, I was an assistant professor who had just finished a PhD on chaos in 2D systems. He, on the other hand, was one of the world's top experts on turbulent mixing. Over the next 30 years, we shared hundreds of long conversations on many topics, both personal and technical. On the personal side, he quickly convinced me to work on applied research - which I have enjoyed throughout my career. On the technical side, on the specific issue of mixing in laminar stirred tanks, he helped me realize that the traditional stirred tank with a concentric impeller has laminar flow patterns that are very close to a 2D system with a rotational symmetry. The key to enhancing mixing under such conditions was to break such symmetries, which we did by varying both temporal and spatial flow patterns. Results were dramatic - and I quickly applied the same ideas to many other systems, including impinging jets, roller bottles, and static mixers.

In this talk, I will review these case studies, and will share many memories of those 30 years of friendship.

399g Fundamental Studies of Laminar Stagnation Flows and Design of Stagnation-Flow Reactors

Triantafillos Mountziaris

This presentation is dedicated to the memory of Professor Robert Brodkey and will focus on fundamental studies of laminar stagnation flow patterns formed by (1) two symmetric counterflowing jets and (2) axisymmetric flows towards a rotating disc. Stagnation flows formed by counterflowing laminar jets are used in a variety of applications, including kinetic studies of diffusion flames, nanoparticle synthesis, microfluidic devices, and polymer processing. Stagnation flows towards rotating discs are commonly encountered in reactors used for vapor phase epitaxy of thin films of semiconductors for microelectronics applications and in analytical instruments used for electrochemical and kinetic studies. Understanding of the structure and stability of these flows is important for the optimal design and operation of process equipment that employ them. Isothermal laminar flows of Newtonian fluids in symmetric counterflowing arrangements exhibit interesting flow patterns above a critical Reynolds number that include time-periodic flows in planar jets and multiple-steady states in axisymmetric jets that can complicate the design of process equipment and devices that employ them [1]. Flow recirculation patterns in rotating-disc reactors used for metalorganic vapor phase epitaxy that are induced by natural convection, disc rotational rate or reactor geometry can have detrimental effects on film quality. Prediction of the onset of flow recirculation is important for optimizing reactor operation. Fundamental studies of the underlying flow patterns can be used to develop generalized flow maps based on dimensionless quantities (Reynolds number, rotational Reynolds number, and Grashof number) and that can guide the design and optimization of commercial rotating-disc reactors [2].

[1] R.P. Pawlowski, A. G. Salinger, J.N. Shadid, and T.J. Mountziaris, Bifurcation and stability analysis of laminar isothermal counterflowing jets, *J. Fluid Mech.* (2006), vol. 551, pp. 117–139.

[2] J. Cho and T.J. Mountziaris, Onset of Flow Recirculation in Vertical Rotating-Disc Chemical Vapor Deposition Reactors, *AIChE J.* (2013), vol. 59, pp. 3530–3538.

AUTHOR INDEX